W9-CYS-018

Sea Slugs

by Valerie J. Weber

Please visit our web site at: www.garethstevens.com
For a free color catalog describing Gareth Stevens Publishing's
list of high-quality books and multimedia programs,
call 1-800-542-2595 (USA) or 1-800-387-3178 (Canada).
Gareth Stevens Publishing's fax: (414) 332-3567.

Library of Congress Cataloging-in-Publication Data available upon request from publisher.
Fax (414) 336-0157 for the attention of the Publishing Records Department.

ISBN 0-8368-4563-3

First published in 2005 by
Gareth Stevens Publishing
A WRC Media Company
330 West Olive Street, Suite 100
Milwaukee, WI 53212 USA

Cover design and page layout: Scott M. Krall
Series editors: JoAnn Early Macken and Mark J. Sachner
Picture Researcher: Diane Laska-Swanke

Photo credits: Cover © Steve Drogin/SeaPics.com; pp. 5, 9, 19 © James D. Watt/SeaPics.com;
pp. 7, 11, 13, 21 © Doug Perrine/SeaPics.com; p. 15 © Andrew J. Martinez/SeaPics.com; p. 17
© David Wrobel/SeaPics.com

Printed in the United States of America

1 2 3 4 5 6 7 8 9 09 08 07 06 05

Front cover: This colorful sea slug rears its head in the
Pacific Ocean near Indonesia. Its "horns" actually help
it smell out food and other nearby nudibranchs.

Table of Contents

Words that appear in the glossary are printed in **boldface**
type the first time they occur in the text.

The Colorful Sea Slug

The butterflies of the sea, sea slugs come in endless styles and colors. Odd bumps shaped like knobs and feathers stick out of their backs. One sea slug has freckles and looks like a squid. A lemon yellow one carries white balloons on its back. Another wears white porcupine needles. A skirt of yellow and orange runs around the side of a silver-white sea slug. Fleshy knobs of dark red cluster on its back.

One of the largest groups of sea slugs is the nudibranch (NOO-duh-brank) family. Between twenty-five hundred and three thousand nudibranchs have been named so far. They range in size from one-half inch (1 centimeter) to nearly 2 feet (61 cm).

Looking like creatures from another world, nudibranchs crawl through our oceans. Divers are always looking for new kinds.

Feathers, Frills, or Fans

Nudibranch means "naked gills." Fish and most other water animals use their gills to take **oxygen** from the water. So do the nudibranchs, but their gills come in very odd shapes.

Some nudibranchs have structures that stick out like fingers on their backs. These structures are called **cerata**. The cerata gather oxygen from the water.

Some nudibranchs carry gills on their backs that look like snowflakes, frills, or fans. Some of these sea slugs can pull their gills into pouches on their backs to protect them.

Tentacles called rhinophores poke out from some sea slugs' heads. Nudibranchs use them to smell for **prey** and to find each other.

Gills shaped like flower buds stick out on this sea slug's back. It is feeding on small colored animals called sea squirts.

Slow Movers

Nudibranchs crawl over the ocean bottom and along **coral reefs**. They can also live in rock pools, cut off from the sea for many hours. They can even **survive** in waters more than 2,300 feet (700 meters) deep. While these sea slugs live in oceans around the world, most prefer warm water in the **tropics**.

Nudibranchs tend to stay in one place or crawl around in search of food. When scared, however, they can swim. A nudibranch called the Spanish dancer bends this way and that to power its way through the ocean.

This Spanish dancer can swim but prefers to crawl along the bottom. Notice its cerata; they look like worms along its back.

Camouflaged with Color

Colors of the rainbow cover many nudibranchs. Many of these sea slugs use those colors to hide. Yellow nudibranchs often feed on yellow sponges, absorbing their prey's color. Some red nudibranchs lay red eggs on red sponges. This **camouflage** helps hide the nudibranchs and their eggs from possible **predators**.

An unusual smell can protect some nudibranchs as well. Some smell like cedarwood when touched. Others carry the odor of sandalwood.

Still others taste bitter. A predator that tries to swallow a sea slug soon learns a lesson. Sea slugs taste nasty!

The orange mop nudibranch on the right in this picture feeds on orange cup corals on the left. This nudibranch's coloring helps it blend into the background.

Colors that Communicate

Other nudibranchs do not bother to hide. Instead, the brighter the colors are, the better. Bright colors send one message to predators — stay away!

Sometimes predators learn the hard way. Many nudibranchs can hurt back. They carry stinging cells on their skin or in the cerata sticking up on their backs.

The Navanax sea slug is one of the few animals that can stand those stinging cells. Streaks of yellow or bright blue swoop down its long, black body. It follows a nudibranch's slime trail and sucks down its prey.

This scrambled egg nudibranch carries poisons on its skin. It is also called the stinky finger nudibranch.

Swiping Cells

Many nudibranchs steal their defenses. Sea anemones and jellyfish carry stinging cells on their tentacles. Most sea creatures would stay away. To touch an anemone or jellyfish is to hurt.

Many nudibranchs, however, nibble on these tentacles. They swallow the stinging cells whole. The cells travel through a nudibranch's body unharmed. Sacs at the end of some nudibranchs' cerata hold the stinging cells. Others keep theirs in pockets on their skin. A nudibranch can lose a cerata and grow a new one anytime.

Two red-gilled nudibranchs feed on a hydroid. A sea animal, the hydroid has a mouth surrounded by tentacles. The nudibranchs use the hydroid's stinging cells in their own defense.

The Chase Is On!

Slowly, the rainbow nudibranch creeps up on the tube anemone. It has terrible eyesight but a terrific sense of smell. Through the water, the nudibranch follows the anemone's odor. The anemone's tentacles wave in the **current**; it suspects nothing.

Suddenly, the rainbow nudibranch shoves its head down the anemone's tube. The anemone pulls its stinging tentacles in quickly, but it is too late. The nudibranch chomps down anyway. Those stinging cells will end up in the nudibranch's cerata.

With a sudden lurch, the nudibranch pulls itself free. Its hunger fed, it twists off into the dark water.

The rainbow nudibranch can ignore the tube-dwelling anemone's stinging tentacles. After all, the anemone will provide food as well as the cells that protect the nudibranch.

A Tongue Like a File

Nudibranchs eat using a horny band called a radula. Rows of tiny teeth grow along the radula. These teeth are made of chitin, a hard substance like your fingernails. Different kinds of nudibranchs have different numbers of rows. They also have different tooth shapes.

Radulas match the kind of prey the nudibranch eats. Some kinds of nudibranchs rake many rows of teeth across a sponge. This gathers bits of the sponge to eat. Other nudibranchs have only a few rows, with only two or three teeth in each row. These sea slugs often have strong jaws. They hold their prey with their jaws and tear it apart with their radulas.

A sea slug munches on a sponge. It has weak jaws but many teeth. The sea slug will absorb the sponge's poisons into its own body. A predator will soon learn how bitter these colorful creatures are.

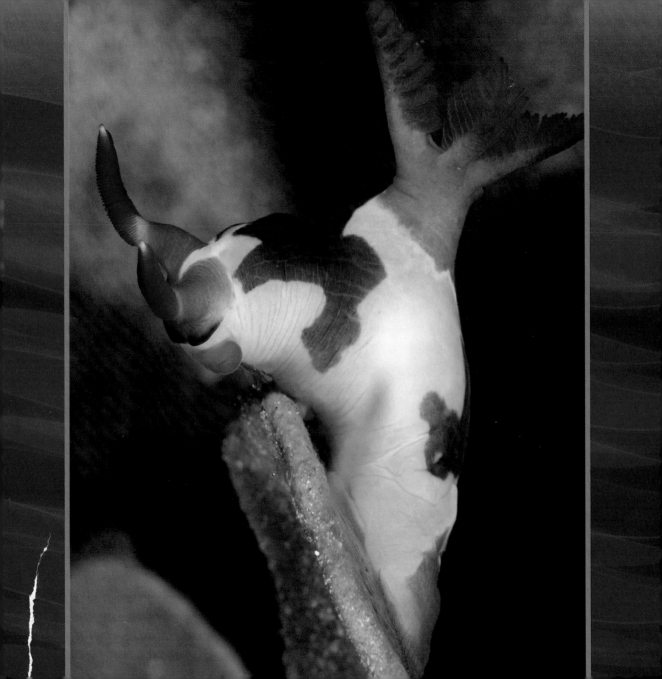

Both Mom and Dad

Most nudibranchs are both male and female. Most start out with both sets of sex **organs**. Others begin as males and become females as they age.

After **mating**, the nudibranch lays its eggs in masses. The egg masses vary in size, color, and shape. Some nudibranchs lay just a few eggs at a time; others can lay thousands. The shape of the egg mass often reveals what kind of nudibranchs will come out of the eggs. For example, the Spanish dancer's egg mass looks like deep pink fabric curled into soft folds. Giving birth to many eggs helps make sure these colorful sea slugs survive.

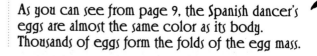

As you can see from page 9, the Spanish dancer's eggs are almost the same color as its body. Thousands of eggs form the folds of the egg mass.

More to Read and View

Books (Nonfiction)   *Coral Reef Feeders.* Marie Bearanger and Eric Ethan (Gareth Stevens)
Journey under the Sea. Linda M. Pitkin (Oxford University Press)
Sea Creatures. Clare Oliver (Raintree Publishers)
Sea Snails. Lynn M. Stone (Rourke Publishing)
Smart Survivors. It's Nature (series). Sneed B. Collard (Creative Publishing International)
What on Earth Is a Nudibranch? Jenny Tesar (Blackbirch Press)

Video   *Secrets of the Ocean Realm: Venom* (Howard Hall Productions)

Places to Write and Visit

Here are three places to contact for more information:

Oregon Zoo
4001 SW Canyon Road
Portland, OR 97221
USA
1-503-226-1561
www.oregonzoo.org

Steinhart Aquarium
California Academy
of Sciences
875 Howard Street
San Francisco, CA
94103-3009 USA
1-414-221-5100
www.calacademy.org/aquarium

**Vancouver Aquarium
Science Centre**
845 Avison Way
Vancouver, BC V6G3E2
Canada
1-604-659-3474
www.vanaqua.org/home

Web Sites

Web sites change frequently, but we believe the following web sites are going to last. You can also use good search engines, such as **Yahooligans!** [www.yahooligans.com] or **Google** [www.google.com], to find more information about sea slugs. Here are some keywords to help you: *nudibranch, rhinophores, Spanish dancer,* and *Navanax.*

oceanlink.island.net/oinfo/nudibranch/ nudibranch.html
Learn about how and where sea slugs live and what they use for defense on this web site.

slugsite.tierranet.com
Click on this site to see the sea slug of the week or of weeks past.

www.diveoz.com.au/nudibranchs/gallery/ default.asp
Twelve pages of these beautiful creatures are displayed on this web site. Click on any image to make it bigger.

www.medslugs.de/E/Med/ Janolus_cristatus_02.htm
This German site presents amazing images of sea slugs from the Mediterranean Sea, the Red Sea, and the Indian Ocean. Click on a picture to enlarge it and get the scientific name.

www.rzuser.uniheidelberg.de/~bu6/ index.html
Find out more about nudibranchs, marine flatworms, and other spineless animals of the ocean.

www.seaslugforum.net
The Australian Museum presents this web site on sea slug behavior and neighboring plants and animals. You can also ask questions of experts on this site.

Glossary

You can find these words on the pages listed. Reading a word in a sentence helps you understand it even better.

camouflage (CAM-uh-flahj) — patterns and colors that make something look like part of its surroundings so it is hard to see 10

cerata (seh-RAT-uh) fleshy tubes on the top surface of nudibranchs that help them breath and digest their food. Cerata come in many different shapes. 6, 12, 14, 16

coral reefs (KORE-uhl REEFS) — ridges or mounds composed of coral, which is made up of the skeletons of tiny sea animals. Coral reefs support a lot of sea life. 8

current (KUR-uhnt) — a part of a body of water that moves along a specific path 16

mating (MATE-ing) — coming together to make babies 20

organs (OR-guns) — parts of the body that do particular jobs 20

oxygen (AHK-suh-jen) — a colorless gas that has no smell. People and animals need oxygen to live. 6

predators (PRED-uh-turz) — animals that hunt other animals for food 10, 12

prey (PRAY) — animals that are hunted by other animals for food 6, 10, 12, 18

survive (suhr-VIVE) — continue to exist 8, 20

tentacles (TEN-tuh-kuhls) — long, thin growths from an animal's body that bend easily 14, 16

tropics (TROP-iks) — a warm area of the world that is near the equator 8

Index

24